Science students are expected to write lab reports, but no book until now showed them how. *Successful Lab Reports* bridges the gap between the many books about writing term papers on the one hand, and the advanced books about writing papers for publication in scientific journals on the other hand. Neither of those groups of books gives much information on science lab reports. *Successful Lab Reports* is designed for undergraduate science students who are beginning to write science. Part I (Format) guides students through the structure as they write a first draft. Part II (Style) shows how to revise the report and polish science writing skills as the student continues to write science lab reports.

SUCCESSFUL LAB REPORTS

Successful Lab Reports

A Manual for Science Students

CHRISTOPHER S. LOBBAN
University of Guam

and

MARIA SCHEFTER
University of Guam

CAMBRIDGE
UNIVERSITY PRESS

Published by the Press Syndicate of the University of Cambridge
The Pitt Building, Trumpington Street, Cambridge CB2 1RP
40 West 20th Street, New York, NY 10011-4211, USA
10 Stamford Road, Oakleigh, Victoria 3166, Australia

First published 1992

Printed in the United States of America

Library of Congress Cataloging-in-Publication Data
Lobban, Christopher S.
Successful lab reports : a manual for science students /
Christopher S. Lobban and Maria Schefter.
p.   cm.
Includes bibliographical references and index.
ISBN 0-521-40404-5. — ISBN 0-521-40741-9 (pbk.)
1. Laboratories.   2. Report writing.   I. Schefter, Maria.   II. Title.
Q183.A1L63   1992
507.2–dc20   91-41418

A catalog record for this book is available from the British Library.

ISBN 0-521-40404-5 hardback
ISBN 0-521-40741-9 paperback

Dedicated to the memory of

JAMES LOBBAN

Father and father-in-law.

Contents

LIKE MAGIC, RIGHT?

Part I Writing the First Draft: Format

Chapter 1 Getting Started

Facing the task

Your first science lab report may seem a daunting task. Whether it's a three-hour lab or field exercise, or an independent research experiment that you have to write up, there are some new skills that you need. Science reports have a format which you must follow. You need to know how to present and interpret data. You must learn to use scientific style in your reports.

This book is your guide to developing the specific skills you need out of your general background in writing. Our goal is to start you writing science. *Successful Lab Reports* bridges the gap between the many books about writing term papers on the one hand, and the advanced books about writing papers for publication in scientific journals on the other hand. Neither of those groups of books gives much information on science lab reports. The chapters in Part I of this book can be used as a tool to help you draft your first report by taking you through the structure of a report. Part II shows you how to revise your paper and refine your scientific style; these are skills you will continue to practice as you write more reports.

Writing a science lab report is not difficult.

No matter whether your experiment went smoothly or seemed chaotic, you can write a good report. Just follow the practical advice in this guidebook. Your grade will probably depend on how well you show that you understand the experiment and how well you write your report using scientific format and style. As you organize your thoughts and data for the report and write simple, clear sentences, you will come to a clearer understanding of the experiment.

Writing for science classes

A lab report is neither a term paper nor a scientific paper. You do not have to read and summarize a large number of books and papers in the library; your report needs only a small amount of background information to give context to your own experiments or observations. On the other hand, while a lab report must have the structure of a scientific paper, it has a different audience and purpose. Your lab report is written to your professor, who already knows all about the work, to show that you understand the process and significance of the experiment. In contrast, a scientific paper is written to fellow scientists to present and discuss new information and ideas.

Good scientific writing is not literary, in spite of the fact that scientists use *literature* as a generic term for their writings. You do not have to be a masterful creative writer to write an excellent lab report! Rather, you need to use straightforward words and clear sentences that unambiguously convey your meaning. The format of a scientific paper already gives you an outline for your report and should get you past the writer's block that can come from staring at a blank page with no idea of what to write.

Structure: the IMRAD formula

The plan of a scientific paper, and so of your report is this:

Title and Author(s)
[Abstract]

Part I: Format

Introduction
Materials and Methods (or Experimental)
Results
Discussion
References (or Literature Cited)
Figures and Tables

The core of this plan is Introduction-Methods-Results-and-Discussion, known as IMRAD by those who like acronyms. This structure arose from the more narrative style of the last century because it helps ever-busier readers find the parts they want.

Read any instructions from your professor!

While the overall IMRAD plan is generally agreed on, writing style is ultimately a matter of individual opinion. Find out what your professor wants. We will indicate by [Ask!] in the following chapters where there are divergent opinions on points of style.

The fact that a paper is presented in the order given above doesn't mean that you have to write it in that order. In fact, if you have difficulty starting with the Introduction, we recommend that you try starting with the Methods. The Introduction is more difficult to write because you have so many ways to begin; in this section, a science report is most like a composition. The Methods section consists simply of factual statements of what you did and what you found. You can use the next few chapters in whatever order you need, but we suggest you read the whole of Part I – on writing the first draft – to get an overview of what you'll be doing. Then re-read each chapter before you begin to write that section. Also, outlining each of your sections will help you to organize your ideas. Writing a lab report is like putting together a jigsaw puzzle: you have to put all the facts and ideas in order, so that they make a clear picture.

6

Outlining

The Introduction and the Discussion do not have much structure imposed by the work. Therefore you must create the structure yourself by preparing an outline. This is very important if you are to avoid rambling. The outline will enable you to create text that is easier for you to write and easier for the reader to follow. Even the more structured sections are easier to write if you first outline them. If you begin writing without an outline and later wish you had one, you can create one from the writing: outline what you wrote to see what you need.

If you use separate pages for each of the sections, you can go back and forth, just brainstorming ideas. When your lists seem complete, organize the ideas. Number the ideas in an order you think will lead the reader clearly from one idea to the next. Any logical sequence of ideas is good; there is no one 'right' or 'best' way to organize a section. Rewrite the lists in order, so you can scan over the structure of each section and think about the flow of ideas. You will probably change the order as you write or when you revise, so there is no need to belabor it in the outline.

Getting started

As you prepare to write, keep one important point in mind: your first draft does not have to be just right, or look pretty. Only you will see this draft – just make sure you will be able to read it later. If you are handwriting, double space and leave wide margins so you have plenty of room for afterthoughts. Don't worry about just the right word or phrase in the first draft; don't bother to check spelling yet, or fix typographic errors. Write – and keep writing! This is a draft and you will revise and proofread it before you hand it in.

Chapter 2   Introduction

Purpose of the Introduction

The Introduction gives the background to the work, starting with the broad context of the study and leading up to the hypothesis – the question studied, or the pattern sought. One way of thinking about the Introduction is as a series of questions relating to why the study is interesting:

How is the organism or process important?
Why was this organism chosen?
   (Convenient to handle? Important 'bug'? Good model system for process under study?)
What methods are available for this study?
Why was your particular method chosen?
What is known to date about the process, especially in this organism?
What question was tested?

The hypothesis

What is the hypothesis? It is a formal statement about nature which can be tested. For example: 'Respiration rate is affected by (is a function of) temperature.' There is a null hypothesis, the opposite, which in this case is, 'Respiration is not a function of temperature.' Or, an ecological example, 'The distribution

9

of organism X on the seashore is (or is not) related to distance above low water line.' The hypothesis need not be written out so formally in the statement of objectives, but the question needs to be clear, and you must answer it in the Discussion.

There is always a hypothesis.

Charles Darwin liked to point out that one could go and count stones in a quarry or grains of sand on a beach, but unless there is a question to answer, such activity is a waste of time and not science.

Even if the hypothesis is not clear in the lab manual or from what the professor has told you, there is a hypothesis, and you must find out or figure out what it is. There can be no experiment without a hypothesis: the whole point of the experiment (or observations) is to provide data with which to assess the hypothesis versus the null hypothesis. When you understand the study, you will know what the hypothesis is. Conversely, if you cannot state the question clearly, you will be unable to answer it; this will indicate to the professor that you do not understand the exercise. Remember that one of the goals of writing up the experiment is to bring you to a clear understanding of it.

Writing the Introduction

For a skillful Introduction, you need to be able to choose relevant facts from the scientific literature, paraphrase them, and reassemble them into your own logical sequence, citing the sources to support each statement. Don't use your lab manual as sole source! Consult your textbook and any references it suggests, or that your professor has put on reserve for you. For a report on a single lab you probably need go no further [Ask!], and your Introduction (and Discussion) should be brief (one or two pages each).

Start broadly, then narrow down to the objectives.

Begin the Introduction by putting the reader in the picture. Let's say your study is on the effect of temperature on yeast respiration. You might start with some broad statement about temperature, or yeast, or respiration: 'Temperature affects overall metabolic rates in organisms because of its effect on chemical reactions. . . . ' Or, 'Yeast (*Saccharomyces cerevisiae*), used in brewing and baking, is sensitive to temperature. . . . ' Or, 'Respiratory rate is a measure of metabolic activity and is affected by many environmental variables, including temperature. . . . ' Then introduce the other broad aspects, using literature as your guide and giving due credit to your sources. For now just write any note that tells you where the idea came from; you can put citations and references in the proper style when you have written the text (see Chapter 10). Keep a list of all the references you have cited.

You may get ideas for an opening line from the first few lines of relevant papers, or from the treatment of the topic in the lab manual or textbook. Sometimes a little history is useful (see the Sample Report in Chapter 11). If you can tie in something practical – biologically, medically, or commercially significant, that will give 'relevance' to your study!

Next, introduce the methods and then the hypothesis. Finally, state the objective. Never write that the objective was to 'learn the technique' or 'to measure this or that.' Whatever pedagogical purpose the professor may have had in assigning this lab, you are writing up the experiment, not the course syllabus. Therefore, the objective is that of the experiment, that is, a question relating the work to the hypothesis.

The outline for the fictional report on yeast respiration might look like this:

1. Yeast – *Saccharomyces* – brewing/baking
2. Importance of temperature
3. Respiration as an indicator of metabolic activity

Part I: Format

4. Methods of measuring respiration
5. Literature on yeast respiration rates, especially as affected by temperature; review articles.
6. Yeast respiration can be conveniently measured in the lab.
7. Relationship between temperature and respiration is a curve with peak at . . .
8. Objective was to use yeast respiration to demonstrate effect of temperature on metabolism.

Chapter 3   Materials and Methods

This section specifies what materials you used and what you did with them. It is one of the easiest to write, especially if you already have a set of procedures written for you. In Chemistry it is often called 'Experimental.' [Ask!]

If you never wrote a science lab report before, you might write a narrative about your adventures in the lab or field, instead of giving a concise account of the materials and the methods. You do not need to include every little detail, but the question is, how much detail do you need? In a published paper, the methods need to have enough detail so that (in theory) someone could repeat the work. How much you need to write for your report depends on whether you are following a lab manual or handout sheet which you can cite, or must write out your entire procedure. [Ask!]

Capsule Materials and Methods

(Read this even if you have to write out the full procedures.)

If you followed a set of written instructions and you are not required to write out the full procedure, you may state briefly what was done and cite the manual:

'We measured photosynthesis of spinach chloroplasts (*Spinacia oleracea*) in a Gilson respirometer, using the procedure in Chen (1970).'

Notice in this example that there is mention both of what was done and what organism you used. The Latin name is a little extra: you don't need it but it will look good – especially if you had to find it somewhere other than in the manual. 'We' in this case would refer to you and your lab partner; don't use 'we' as a substitute for 'I'. Don't just write, 'We did the experiment like the book said.'

Suppose you didn't exactly follow the written procedures. This will be especially likely if you used a published lab manual, because the professor will probably have adapted it in some way, such as by using a different organism or slightly different apparatus. In addition, you may have changed the procedure (deliberately or inadvertently). If there were modifications, you should specify how your actual procedure was significantly different.

What is significant? This is a tricky question; the easy answer is, anything that might have made a difference to the outcome. [See the boxed story on page 18.] For instance, if the experiment called for NaCl but your instructor gave you KCl instead, this is worth noting. So is a 0.5 M solution instead of 0.2 M. But if you had a 250 ml flask instead of a 125 ml flask, perhaps that makes no difference. After you consider sources of error (Chapter 6), you will have a better idea of what to include or exclude, but for now make notes of what you think is appropriate. You can add or subtract later. Remember, this is a first draft.

Writing the complete Materials and Methods

Start with an outline. You need to answer questions such as these:

What organism was used (give species name; if a culture, a strain number may be needed)?

Part I: Format

Methods: A Case History

Here's a story to illustrate how difficult it can be sometimes to know what is 'significant' in materials and methods.

About the turn of the century, a German cytologist named Martin Heidenhain developed the hematoxylin stain for chromosomes, which is still in routine use today. He had beautiful results which everyone wanted to copy. But try as they might, others could not get the stain to work. In spite of many letters back and forth, no one could discover the problem. Finally Heidenhain, who was as puzzled as they were, invited a delegation to his laboratory in Würzburg. They discussed their attempts, while Heidenhain twirled his moustache with his stained fingers. Finally he said, 'So, you will watch me, please.' And he began at the very beginning to prepare the stain.

Everything was exactly as written until the venerable professor picked up a rusty spoon to stir the solution. He thought nothing of it, but his guests remembered that iron is used by weavers as a mordant to make their dyes bind, and they realized that this was the missing step. So now iron is added to the stain solution. Isn't science lucky that Heidenhain didn't stir the solution with a wooden pencil?

What parts of the organism, how big? If cultures, how much or how dense?

How many replicates?

In the field, where was the study site? How big was the sample size? How were organisms harvested, weighed, etc.?

What apparatus or equipment was used?

What were the experimental conditions (temperature, light, medium, etc., as appropriate)?

What was the composition of the reaction mixture?

How long were the treatments?

What were the controls?

How were data handled (calculations, statistical tests)?

18

Remember that you are writing the report to your professor, not providing a set of instructions to another student. You can leave out certain steps as obvious (e.g., that the test tubes were clean, or that [when you inoculated a flask] you took out the stopper and then put it back). You lab manual may include many such extra details for your benefit, but you do not need to repeat them in your report.

Don't keep using 'then'. If you write out a series of steps, the reader will understand that they were in that order (therefore put them in order!). 'Then' is OK now and then.

Chapter 4 Results: I. Graphs and Tables

The first step in writing the Results is to draw whatever graphs and tables you need from the numerical data you recorded, and to do any calculations necessary. Then you will be able to make statements about what you found. Data presentation is the core of scientific reporting and this is why we emphasize it.

Table, graph, or text?

Clear graphs and concise, intelligible tables are as important as clear, concise writing. If your manual has not specified how you are to present your data, and you cannot state the results simply in a sentence, you must choose a table or graph – but not both for the same data. Graphs are generally more easily understood; any data that show a trend should be graphed. If there is no trend, or if exact numbers are more important, use a table. Good examples are shown in Figures 1 and 2 and in Table 1 (page 32).

The type of graph you draw will depend on the kind of data you have. Continuous data, such as the relationship between respiration and temperature, need a line graph (Figure 1), whereas discrete or classified data, such as the numbers of individuals in different size or age classes in a population, must be drawn as a histogram (Figure 2).

Part I: Format

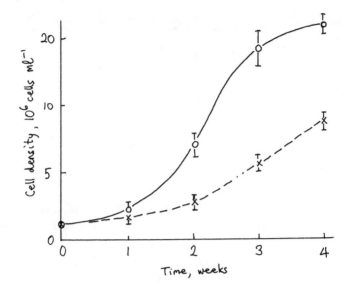

Figure --. Growth of species Z in culture
with (o—o) and without (x--x) added
nutrients.

Figure 1. Example of a properly designed figure (fictitious data).

A good histogram is a graph, with a scale on each axis. If the *x* axis is a list of unrelated items, then the figure no longer shows trends and is merely a chart.

Data such as percent composition can often be stated simply in a sentence. This may lack the visual impact of a pie chart (some would call it boring), but it takes up much less space and is easier for you to do. For instance, 'Cell dry matter was 20% protein, 10% fat, 30% carbohydrate, and 40% ash.' If you have more such data – the compositions of five species, for instance – make a table.

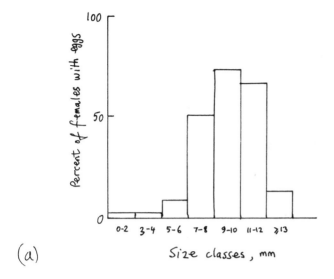

(a)

Figure --.   Percentage   of   gravid   female
copepods in plankton tows from outer estuary,
April 6.

Figure 2. Example of a well designed histogram with appropriate data (fictitious).

Graphsmanship

The following suggestions assume that you will draw your graphs by hand. Even if you're going to do your (final) graphs on a computer, you need to know how to design them, and it's a good idea to sketch them by hand now to see what you've got. If you are familiar with a computer graphics program, you can certainly produce some very smart graphs. But if you're not very familiar with the program, beware: you can waste a lot of time trying to get labels and axes right (and maybe even lose your data!).

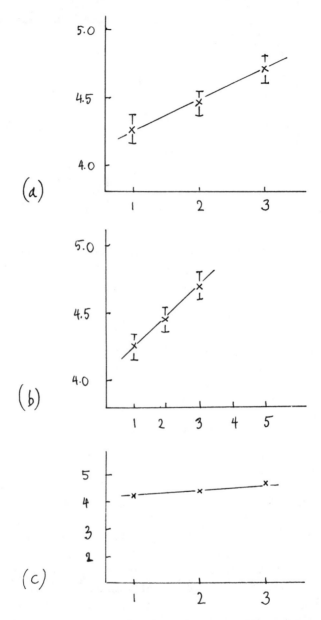

Figure 3. The same data can appear to show a steep increase, a moderate increase, or virtually no increase, depending on the scale used. The data in all three graphs are: 1 = 4.25±0.10; 2 = 4.45±0.10; 3 = 4.70±0.10.

Plan your axes: What range of x and y do you need to cover? (Remember the dependent variable – the measured response – goes on the y axis, and the independent variable – the one you controlled – on the x axis.) How many squares will you need? Check the largest and smallest numbers you must include. (The smallest number will often but not always be zero.)

The interpretation of the graph may change depending on the scale – you can overemphasize or obscure trends by changing the scale (as advertising agents well know!). Figure 3 illustrates the point. If the theoretical values range from 4.0 to 5.0 (as in Fig. 3a), these data suggest a strong relationship. If the theoretical range were 0–10 (Fig. 3c), the data suggest there is no relationship. A real example is given in the box on pages 26–27.

You may not be able to tell in advance what scale to use, but if you don't get the graph you expected, consider the scale on one or both axes. Obviously, you must have some basis for comparison.

Axes marks should be 'round' numbers – 0, 5, 10 or 0, 2, 4 – even if the independent variable was set at numbers such as 8, 11, 16 or 2.5, 3.7. Your data points do not have to be on major lines! A graph with an axis labeled 8, 11, 16, 21 is very hard to read. When we look at a graph we often estimate intermediate values, so make it easy.

When you have planned the size, draw the axis lines and mark the intervals. Always draw in and label both axes. Don't just use the edge of the grid. Actually draw the axis lines. In fact, the margins on most looseleaf graph paper are irritatingly narrow for writing in the labels (you also need a caption), so it's better to make the graph smaller than the whole page.

You must label both axes clearly with what is shown and the units, for example, Respiration, μg O_2 (mg dry wt)$^{-1}$ h^{-1}. Plan enough room below or to the left of the axis so you won't have to crowd the label onto the interval numbers – don't be cramped by the narrow margins. Center the label as best you can.

How many lines on one graph? Putting two or more related

25

The Case of the Missing Buffer Zone

A student group had titrated an organic acid. Theory says there is a buffer region, and the theoretical curve looks like Figure 4a, with a buffer region in the middle. Notice that the *x* axis is given in equivalents of base. The students titrated and recorded the data for amount of base directly in milliliters; this should have given the same curve. The manual told them to titrate until there was no further change in pH, and they kept going a bit further to make sure. Then they plotted their data as pH versus ml of base added and got the graph in Figure 4b. Unfortunately, it didn't resemble the theory. The ascending part of the curve might have had a minor dip in the middle, but given the small errors in the data points, the students couldn't be sure. Also, there was an unexpected plateau at the end. The problem was that they had gone on adding base too long, but this was not obvious until the amount was calculated as equivalents. Beyond the end point of the titration there is no pH change because the solution is swamped with base. When they cut off the plateau and stretched out the *x* axis, the buffer region suddenly appeared.

---

curves on the same graph can help highlight a comparison, for example, between parallel treatments or the same treatment on two organisms. The lines must be clear, however. If they cross over too much or get too crowded, the graph becomes confusing – worse, not better, than separate graphs. If you put more than one set of data on a graph, use different symbols (squares versus triangles, or filled versus open symbols) and draw different lines (e.g., solid versus broken – not different colors) (see Figure 1).

Make the data points big enough to be seen clearly – at least 1 mm diameter, not just a pinpoint on the grid. Not only should you be able to find the points to draw the line through them,

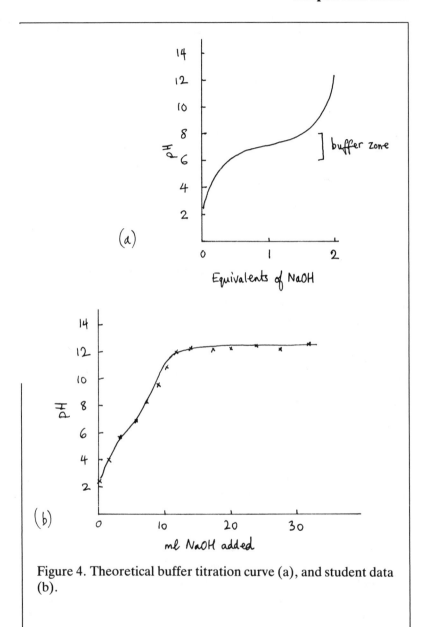

Figure 4. Theoretical buffer titration curve (a), and student data (b).

the reader should be able to see them, too. Making a tiny dot for 'accuracy' involves the same illusion as having too many significant figures in your calculations. Your data are not extremely precise and that's OK – there is an inherent error in all biological work. Standard deviation can be calculated on replicates and shown as error bars on the independent variable (see Figures 1 and 3). There is also (but rarely shown) error in the value of the *dependent* variable. (For instance, was the temperature exactly 20° C all through the experiment or did it vary a bit?) Readers will assume that the center of the symbol is on the very point you mean, so don't be afraid to draw big symbols (2 mm).

Now you can draw the curve. Most experiments in practical courses are designed to illustrate some relationship, so you should try to show it. Probably you should not just join the dots. For instance, a calibration curve will usually be a straight line and you need to draw it that way, ignoring the small experimental errors, so that you can accurately interpolate to find the value for your unknown. Think of the line as an interpretation of the data. The interpretation that you try (at least initially) depends on the theory: consider the two graphs in Figure 5.

In many scientific papers, where the data are new and may or may not fit some existing theory, the data points (means) are joined by straight lines. If you don't have a theory to work with, or your data are contrary, then join the dots.

Finally, the figure needs a caption (also called a legend) to say what it shows. Give enough information to explain what is shown (e.g., which experiment, what the symbols mean if you used more than one). The caption is not simply a title; see the example in Figure 1. Put the caption underneath the figure if you have room. If you used the full sheet of graph paper you can put the caption on some bare part of the graph, such as the upper left or lower right.

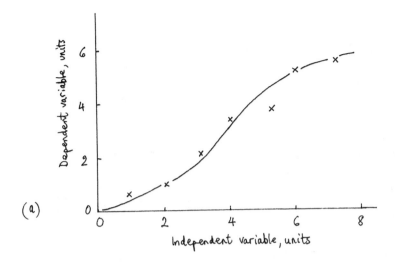

(a)

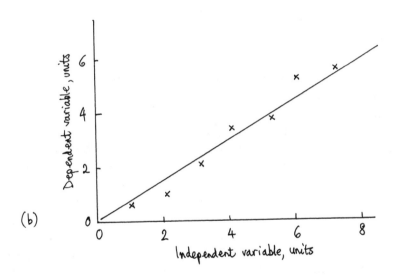

(b)

Figure 5. The same set of data subject to two interpretations: (a) supposes a sigmoid curve, such as a growth curve; (b) supposes a straight line, such as a calibration curve.

Tables

Any data that are inappropriate as a graph and cannot be reduced to a single statement in the text are presented as a table. Don't repeat any data that are in graphs. Tables, like figures, are to be prepared on separate pages, not mixed with the text. Type tables double spaced, just like text.

Each table needs a title at the top. In contrast to figures, for which all the writing goes in one block, explanations and comments for tables are added as footnotes, cued by letters, numbers, or symbols (see Table 1). This is one of very few places in science writing where footnotes are acceptable.

Plan the layout of your table. Comparable data should be arranged in columns, to be read down, rather than in rows. Columns are much easier to read than rows. Have columns only for informative data – you don't need a column that is all the same number (especially zeros). For instance, if temperature was the same in all cases, simply give the experimental temperature in the caption (or Methods). Plan the spacing of the columns to allow for the material in them, as well as the space needed to write the column heading, and the number of columns you have to fit across the page. Turn the paper 90° counterclockwise if you need more room (the top of the table toward the binding edge of the paper).

Make the column headings as brief as possible. You can use abbreviations; define them in a footnote if they are not obvious (e.g., Temp for temperature). Most of your columns will have numbers in them, which don't take up much space. Keep your table compact by having short headings. Units must be specified in the column heading.

Draw lines below the title, under the column headings, and below the columns of data. Do not use other horizontal lines – if you want to separate groups of data, leave an extra space – and do not use vertical lines between the columns.

Figures and tables should be collected as two groups either

Part I: Format

Table 1. Effects of antimicrobial compounds against two bacteria in the Kirby-Bauer test.

Antibiotic	Bacillus sp.		E. coli	
	zone [a]	sensit. [b]	zone	sensit.
Ampicillin	11±1	R	19±1	S
Chloramphenicol	24±1	S	28±4	S
Erythromycin	28±3	S	15±4	I
Gentamicin	20±2	S	19±1	S
Nalidixic acid	17±2	R	14±6	I
Penicillin G	8±0	R	13±7	R
Polymixin B	11±6	R	13±2	S
Streptomycin	19±4	S	14±5	S

[a]Mean width of inhibition zones to nearest mm ± SD.

[b]Sensitivity based on comparison with Kirby-Bauer chart (Benson, 1985): R = resistant, S = sensitive, I = intermediate.

Table 1. Example of a properly formatted table, taken from "A. Student's" revised lab report (Chapter 11).

at the end of report (standard for submitting manuscripts for publication) or in the Results section. Don't try to integrate text and illustrations on the same page. It's much easier to type your text through without having to plan spaces for figures and tables.

Chapter 5   Results: II. Writing the Text

When you have sketched out the graphs and tables, you are ready to draft the Results section. Look at the data and decide what they say. The structure of the section is largely dictated by the structure of the work – experiments 1, 2, 3, and so forth.

You must make some statements about your results.

State the results briefly without describing the curves or repeating the data in the tables. Start with the most important result or observation, if you have several. Describe the overall results, not each separate measurement, except where there were unusual data points. Remember to use the past tense. Write something about each graph or table, and refer to each parenthetically. For example, 'Photosynthesis was saturated at a light intensity of 250 μE m^{-2} s^{-1} (Figure 1).' Graphs and tables present data, they do not state results, so do not simply offer the data as your results (as in, 'The data are presented in Table 1 and Figures 1, 2.'). If you later find yourself introducing results into the Discussion or discussing results that you neglected to present, insert the data into Results.

Do not include references to other works (published data or statements of theory). Give only your results. The place for comparing your data with theory and for interpreting them is in the Discussion. 'Your' in this context means you and your

Part I: Format

lab partner, if any, or even the whole class if the collective data are in your report.

As you write, number your graphs (and any drawings such as maps or experimental apparatus) as Figures, in the order in which you refer to them. Number the tables, in order, with a separate series of numbers.

The Results should state only what you found.

When you have made your statements – STOP! Hold the interpretation for the Discussion. Look back over your draft Results, checking against the graphs and tables. Did you miss anything?

Chapter 6　Assessing the Results: Notes for the Discussion

Before you can write or even outline the Discussion, you need to assess the results, taking problems and errors into account. To do this, you need some basis for comparison: what did the hypothesis predict? What have other people found? If the results were not as expected, what are the possible reasons? Finally, you must consider whether the hypothesis or the null hypothesis (or neither!) is supported.

More about graphsmanship

In comparing your results to the literature, you must look closely at the published results and the experimental conditions. Consider the two graphs in Figure 6. Do they show the same relationship between respiration and temperature?

At first glance, the two graphs seem quite different: published data (and theory) show a U-shaped curve (Fig. 6a), the student found a hyperbolic curve (Fig. 6b). The student's axis scales seem fine, as far as her own data are concerned. The difference lies in the range of temperatures tested. The student's experiment included temperatures only up to 25° C; the decline in respiration occurs above that. Her data are consistent with the hypothesis as far as they go.

What if the two graphs proved to be actually different? Next you need to look at the experimental conditions. Was the same organism used (species, strain)? Were other environmental

Part I: Format

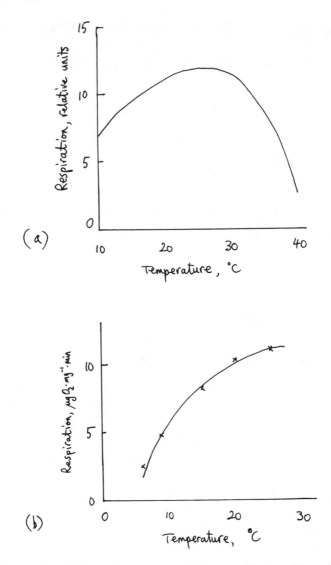

Figure 6. Textbook curve (a) and student data (b) for effect of temperature on respiration.

conditions the same (nutrient status, culture medium, oxygen concentration, pH, etc.)? Many factors can affect the response, especially quantitatively. (For example, at a different salinity, a peak in respiration may occur at a lower rate or at a higher temperature.)

Sources of error

Your data will have various errors which you need to evaluate. The objective is to discover the reliability of your data, the extent to which they can be trusted to tell you about scientific reality. To pick an extreme example, someone might take a set of readings that were all zero . . . because the instrument was not switched on. Discovering the sources of error is like trouble-shooting a machine or a computer program, except that you cannot go back to the experiment, make adjustments and try again. You are analyzing after the fact.

There is always some error in biological measurements and experiments; that is why replicates and statistical samples are so important. It is also why you do not join the dots in calibration curves.

Accounting for the sources of errors has nothing to do with assigning blame.

Do not blame yourself or others for what went wrong. Simply consider the sources of error and the extent to which they affect the data.

Here are some potential sources of error:

Experimental errors: for instance, adding a bit more or less of a test solution; or accidentally losing some biomass during dry weight determinations. Such errors will displace the data point along one or the other axis. No matter how precise you think you were, there will always be some variation. There's little you can do about these errors after the fact. If you know that you lost half your sample, you should discount the data

from it! More likely, you did not notice the slip, or the error was not that great, and you must include it.

Sampling errors: in a variable population, whether of organisms or of test tube replicates, smaller samples (fewer replicates) are more likely to miss the real mean value.

Errors in measurement: it is easy to misread an instrument especially if it has analog read-out, such as the needle on a spectrophotometer. When the value is changing with time, you have to try to pick a value and simultaneously note the time. You can easily be off a bit. Also, you may have introduced some unnoticed bias. If the number was hard to read, you may have erred in favor of the theory – what you (subconsciously) thought the number ought to be.

Errors in recording or recopying: the number that you record should always be the same as the number you read, but sometimes your brain will skid. Maybe you'll write 65 instead of 56. Occasionally the error may be big enough that you can spot it ... but you probably can't fix it. For instance, a pH of 41 is impossible. But was it really 14 or 4.1? If you can't be absolutely sure, don't guess. You can reduce this source of error by developing the habit of double-checking numbers at the time you record them. Also recheck when you copy from your lab notes onto tables and graphs.

Errors in computation: double-check your calculations; cross-check if possible. Don't assume that your classmates had it right if your answer is different – YOU may have the correct answer! Go back to the raw data if you need to. Common errors in computation include mis-entering a number into the calculator/computer and misplacing a decimal, especially when converting between units (e.g., milligrams to micrograms). Make sure your units are consistent. The more complex the calculation, the less likely you can eyeball a mistake, but you should think about whether the answer is reasonable. If you add 5 and 27 on a calculator and get 12, you know you mis-hit something; in this case apparently the 2 was not entered (or perhaps the battery needs replacing). Look at a mean to see if it looks right – you may have hit "enter" without putting in data, and so created a spurious zero datum. (That can have an effect like one C on a grade point average!) Compare your results with expected values to see if you are close.

Differences from published procedures: Any change in the procedures, whether intentional or inadvertent, by the professor or you, has the potential to create differences ('errors'?) in the results as compared with theory. Look over your Materials and Methods and lab notes to see what changes there were.

List the changes and sources of error and try to decide for each if it would have a major or a minor effect on the results – or no significant effect – and in which direction the results would change, if at all. You need to know something about the process studied to make such judgments. Obviously, if you accidentally killed the organism before the experiment, it wouldn't have performed as expected – big factor! But if you substituted one strain of the organism for another, perhaps there would be little or no effect. When you've decided which changes might have been significant, list them in the outline for the Discussion.

Chapter 7   Discussion

The Discussion answers the question, What do the results mean? It is essentially an argument about the hypothesis based on the results. Since all the experiments or observations (presumably!) relate to the hypothesis in question, you must in the end draw your conclusion as to whether the hypothesis or the null hypothesis is supported. Although you could discuss results 1, 2, 3 in order, you will find that many comments either refer to several results or compare two or more results. You can more usefully organize the Discussion into a number of smaller questions:

1. What is your interpretation of the results, in light of the hypothesis and the published literature?
2. What are the significant sources of error in the results?
3. Therefore, how reliable are the results?
4. Do the results support the hypothesis or the null hypothesis? (Or perhaps neither or both?)
5. What changes in procedure would give better results; what additional experiments would help support or refute the hypothesis?

Remember that whether your experiment worked as planned or deteriorated into complete chaos, you can write a good report. The key section in the report is the Discussion, because it is

where you show that you understand what should have happened and what did happen.

You should start the Discussion by stating your interpretation of the facts, in as positive a way as you can (without exaggerating!), perhaps comparing or contrasting them to the literature. Only then should you tackle reliability and errors. Never start the Discussion by lamenting what went wrong; consider the strengths of your data first. For instance, 'The respiration curve, with a peak at 28° C, was similar to the results of Martinez (1978).' Or, 'The respiration curve showed a clear peak at 28° C, in contrast to the theoretical curve given by Martinez (1978).'

You should freely admit a failure to get results; show that you understand what might have gone wrong. Notice that there is an important difference between negative results and a failure to get results. In a teaching lab you can be fairly sure that the principle you have been given to demonstrate is a well-established hypothesis, so if your results don't demonstrate it, something went wrong, that is, you failed to get results. You could report, 'I was unable to show a relationship between respiration rate and temperature.' You could not conclude that, 'There was no relationship between respiration rate and temperature.'

The point of examining the errors is to evaluate the results. Do not assign blame or assume any guilt . . . even if you or your lab partner did something really stupid! Give a cool, scientific explanation.

Finish the Discussion by drawing your conclusions and relating them to the Introduction. Don't make grandiose claims for a modest experiment. You don't have to pretend that your experiment or observations advanced science, and it looks silly if you do. And don't conclude that what happened was that you learned how (or how not!) to do the experiment. Address the hypothesis. If appropriate, suggest how to improve the experiment, or what additional experiments would be helpful.

Leave the reader with a positive message. This is the end of the text of your report, and you want your professor to feel that the report is complete and satisfying. The worst possible way to end is just to stop at the end of a litany of errors. That way you leave the reader thinking that the lab work and the report were a mess. One good way to end is to return to an idea in the Introduction, especially one relating to the hypothesis.

Chapter 8 Abstract

One last section you may need for your report is an abstract [Ask!]. The abstract is an important part of a published paper in these days of electronic retrieval because it (and the title) may be all that most readers see of the paper. Even 'Notes' (short scientific papers), which often used to lack an abstract on the assumption that something short doesn't need summarizing, now usually have one. Your lab report is not going to be published, and you can be sure your instructor will read it all, so perhaps it does not need an abstract. However, your professor may wish you to practice the skill of writing an abstract. If so, read on.

The abstract is a self-contained synopsis of the report.

The Abstract is not a 'hook' to induce the reader to keep reading. So it is unlike the first paragraph of a magazine article or English composition. Nor is it like a blurb for a book or movie, leaving the audience to anticipate the outcome. Finally, the Abstract is not the same as the 'Summary' that sometimes appears at the end of a scientific paper. Such summaries cover only the conclusions.

The emphasis in an abstract is on the results and conclusions. It should have only the objectives from the Introduction, and only a brief reference to the Materials and Methods (unless the experiments focused on methods). It summarizes results

47

(actual values) and conclusions; it may state the hypothesis that was supported (or refuted). Since the Abstract must be completely self-contained, it cannot include any literature citations or references to your own graphs or tables.

The Abstract must be informative.

Contrast these two abstracts for a report on enzyme kinetics:

Good:
Enzyme Xase in cell-free extracts of Species Y showed V_{max} of ... and K_m of ... The hyperbolic curve is consistent with Michaelis-Menten kinetics typical of non-regulatory enzymes.

Poor:
Activity of enzyme Xase was measured and the results discussed.

Part II Crafting the Final Version: Style

Chapter 9   Revising Your Paper

Looking back over the whole first draft

If you have worked through Part I, you now have a complete text for your report, either handwritten or printed out from your computer. It is in the proper format but the style may be rough. Set this draft aside for a day or two, so that you can come back to it with a critical eye. (If you cannot wait that long, at least do something else for a few hours.)

The way you go about revising depends partly on whether you are working on a PC or have been writing by hand. If you are writing on a PC, edit your printout – don't work on the disk yet. You can spread out the pages and scan from one to another much more easily than you can on the screen. Mark changes on the hard copy to type in later.

However, if you have a handwritten draft, it may need typing so that you can easily read it and will have room to make changes. (If you haven't yet found access to a PC, revising will give you plenty of incentive!) Also, your writing seems better in your own handwriting than it does typewritten; you'll be able to revise the typescript more easily. When you type, think about details of grammar and style (see below), rather than thinking about the whole paper. Typing forces you to look at your writing word by word. If a word or phrase seems awkward to you, it probably is. Change it if a better expression comes immediately to mind; otherwise, just mark it to think about later.

Improving the content

First read through the report to get the big picture. Consciously think about large aspects. For a start, does it have all its parts in the right order? Flip through and check them off:

Title and your name
Abstract (if needed)
Introduction (with objectives)
Materials and Methods
Results (only your results)
Discussion (with a conclusion)
Acknowledgments (if needed)
References
Tables (numbered and with titles)
Figures (numbered and with captions)

Next go over the structure carefully:
Is each section complete? If you were shifting back and forth between sections or were hurrying, you may have left out something. Try not to overlook the obvious.

Do the sequences of ideas in the Introduction and Discussion flow, with smooth transitions between them? Ask yourself at the start of each paragraph, 'Does this follow from the last paragraph?'

Have you given enough background in the Introduction so that the rationale for the objectives is clear?

Does the Discussion evaluate the results and give the conclusion about the hypothesis you tested?

Are the graphs and tables complete and properly designed? Is each cited in order? Did you avoid duplication? Could you reduce any to a sentence?

Results versus Discussion

One of the finer points of scientific style is the separation of Results and Discussion. The Results section should include only what you found and the Discussion what that means, but the distinction between the results themselves and their interpretation is not always sharp. Sometimes the way you write a sentence will convert a result into a conclusion. Recall, for instance, the way of reporting a failure to get results (p. 43). Consider why the following phrasing is a conclusion, not a result:

'Temperature had no effect on respiration.'

What you observed was that, 'Respiration rate was a ± b at all temperatures tested.' The conclusion includes an assumption that no other factor masked any effect of temperature.

Statements of values will usually give only results; statements that involve comments on cause and effect are probably conclusions. Another clue to discussion in the Results is any reference to the literature. The Results should contain only your data.

While the difference between results and discussion may seem a small point in writing, it bears on an important scientific issue. We don't always distinguish between observations and conclusions in everyday life – we often see only part of an event and fill in the rest. We could not operate if our brains did not function this way. Safety and survival depend on quick assumptions (red stove elements are hot!), but as scientists (and even in everyday life) we need to be able to recognize the difference between observations and conclusions. (See box.)

Improving the style

Whenever you read through your paper you should be alert for grammatical or spelling errors. In addition, you should read through the paper once to consider the writing carefully and to give it a final polish. However, the main thing is to present your scientific work clearly. You need not belabor style, particularly composition style, by trying to think of more elegant

---

Sherlock Holmes, the Master of Assumption

"Doctor Watson, Mr. Sherlock Holmes," said Stamford, introducing us.

"How are you?" he said cordially, gripping my hand. "You have been in Afghanistan, I perceive."

"How on earth did you know that?" I asked in astonishment.

"Observation with me is second nature," he observed. "From long habit the train of thoughts ran so swiftly through my mind that I arrived at the conclusion without being conscious of intermediate steps. There were such steps, however. The train of reasoning ran, 'Here is a gentleman of the medical type, but with the air of a military man. Clearly an army doctor, then. He has just come from the tropics, for his face is dark, and that is not the natural tint of his skin, for his wrists are fair. He has undergone hardship and sickness, as his haggard face says clearly. His left arm has been injured. He holds it in a stiff and unnatural manner. Where in the tropics [in 1880] could an English army doctor have seen much hardship and got his arm wounded? Clearly in Afghanistan.' "

"It is simple enough as you explain it," I exclaimed.

"I am an expert at jumping to conclusions," he declared.

[Adapted from "A Study in Scarlet" by Arthur Conan Doyle.]

---

ways to express yourself. Check spelling and basic grammar, and allow your sense of style to develop gradually.

Three points of scientific style need special mention. These are: conciseness, tense, and the use of the passive voice.

1. Scientific writing is concise.

Concise means expressing much briefly – it is not simply short. Never sacrifice clarity for the sake of brevity. If the sentence becomes so condensed that the reader has to fill in words to make sense of it, there is a great risk of ambiguity. This is why headlines or classified ads are sometimes funny. Eliminate repetition and wordiness but check that every sentence is unambiguous.

2. The question of tense

You need to recast instructions written in present tense or imperative (Do this, do that) into *past tense* (This was done, we did that). The experiment is over and is to be written up in the past tense. However, scientific etiquette requires that when you cite a published result, you refer to it as a fact which still *is*, even though the result was found in the past: 'DNA has a double helix structure (Watson & Crick 1953).' However, if you write about the discovery, use the past tense: 'Watson & Crick (1953) discovered the double helix structure of DNA.'

3. Active or passive voice?

The passive voice makes the object of the action the subject of the sentence: 'Five bottles were filled.' The real subject – assumed to be you, as author of the report – is omitted. Science papers are often written in the passive voice. However, there are divergent and often strong opinions about whether use of the passive is good or bad. [Ask what you professor wants!] One of our colleagues aggressively promotes the passive. ("The report is about the experiment, not about you!" she exclaims.) Another colleague equally hotly denounces use of the passive as "Victorian prudery" that leads to "committee writing [style], ponderous discussions, and avoidance of responsibility." The passive is generally cumbersome, wordy, and dull; unless skillfully written, passive sentences may be confusing or even pompous. The passive voice may have a place in the Methods section but we suggest you use it sparingly and cautiously. You will find the active voice easier to write, since that is how you learned to write, and your sentences will be clearer and more direct if you use active verbs.

Check singulars and plurals, especially of words with Latin or Greek origins: Data are plural. So are media (singular = medium), mitochondria (mitochondrion), criteria (criterion), and flagella (flagellum). But not all Latin-looking words ending in -a are plural! Examples include alga (plural = algae), seta (setae), stoma (stomata), and stigma (stigmata), among others.

Think PARAGRAPHS

Group together any notes that relate to the same idea and put them in the same paragraph when you write. A paragraph has several sentences relating to a single theme. The theme is often stated in the first ('topic') sentence. Do not write single-sentence paragraphs. Do not mix themes within one paragraph.

There are many guides to proper usage of English, including some for sciences writers (see the Further Reading list on pages 101–102). Below we summarize a few points drawn from Woodford (1968) and Day (1988).

The Council of Biology Editors' book *Scientific Writing for Graduate Students* (Woodford 1968) gives rules, with exercises on each, for polishing style. We have excerpted the following, with permission:

Make sure of the meaning of every word, especially pairs like varying/various; affect/effect. There are lists in the *CBE Style Manual* and in Day (1988). The context of the word may be as important as the meaning of the word itself. For instance:

'The tubes were shaken, followed by centrifugation, and the upper phase withdrawn.'

(Were the tubes followed by centrifugation? Were the upper phase withdrawn?)

Use verbs instead of abstract nouns, for example, *separate* rather than *separation*. Releasing the hidden verb makes the sentence shorter and more vigorous:

'Primary and secondary particle separation was obtained by electrophoresis.'

becomes:

'Primary and secondary particles were separated by electrophoresis.'

Break up noun clusters and stacked modifiers; your sentence may be longer, but it will be clearer. The longer the string, the less intelligible it becomes. For instance:

'Highly purified heavy beef heart mitochondria protein . . .'

This is *too* concise – what modifies what? In this case the meaning is:

'Protein from the highly purified heavy fraction of bovine heart mitochondria . . .'

Day (1988) gives Ten Commandments of writing, humorously incorporating the fault into the commandment:

DAY'S TEN COMMANDMENTS OF GOOD WRITING

1. Each pronoun should agree with their antecedent.
2. Just between you and I, case is important.
3. A preposition is a poor word to end a sentence with.
4. Verbs has to agree with their subject.
5. Don't use no double negatives.
6. Remember to never split an infinitive.
7. When dangling, don't use participles.

8. Join clauses good, like a conjunction should.
9. Don't write a run-on sentence it is difficult when you got to punctuate it so it makes sense when the reader reads what you wrote.
10. About sentence fragments.

The many books about term papers cover these points of general style, and also give more tips on paragraphing, paraphrasing, and sentence structure. The Further Reading list gives references to several of these, as well as the more advanced books on scientific writing.

When you have polished the wording of your report to your satisfaction, you are ready to 'package' it. The finishing details are:

1. Put citations in the proper style and complete the reference list (see Chapter 10 for what you need);
2. Draw the final graphs;
3. Print out or type the final copy; double space everything, including references and tables;
4. Proofread one last time for typographical errors . . .

. . . and hand in your report. Congratulations!

Chapter 10 Citations and Reference List

Scientists who write papers for publication have learned to deal with the exacting details of citation and references style demanded by journals. Your professor may wish you to begin learning the style now. He or she may be more concerned, however, that you use the literature and write a good report than that you write out references in some precise format. [Ask!] Correct information is more important than correct arrangement of the items.

Check the information carefully!

Make sure you copy the references correctly, especially the volume number and page numbers. As you use books and papers to track down others you will quickly learn that a mistake in a reference can be extremely frustrating.

The first step in finishing the reference list is to check through your report and list all works that you cited. Do not create a bibliography, that is, a list of pertinent works which you consulted but did not cite.

The list of references follows the Discussion (and Acknowledgments, if any). The references must be arranged in alphabetical order of the names of the first authors (who are also sometimes called the 'senior' authors regardless of their relative ages!).

Indent the second and subsequent lines of each reference

Part II: Style

(see samples below). Do not start a new line for the date or title (in contrast to the style common in the humanities).

Journal article:

Milligan, C.L. & C.M. Wood. 1987. Muscle and liver intracellular acid-base and metabolite status after strenuous activity in the inactive, benthic starry flounder *Platichthys stellatus*. Physiol. Zool. 60: 54–68.

Book:

Day, R.A. 1988. How to Write and Publish a Scientific Paper, 3rd ed. Oryx Press, Phoenix, AZ.

Citing scientific literature

Scientists pride themselves on making discoveries – discovery is the essence of science, so giving credit to other people when you cite their findings is very important. Scandals and lawsuits have resulted when one scientist appeared to adopt the work of another as his own. On a lesser scale, taking other people's work without giving credit is plagiarism, a most serious academic offense. Your name on the paper implies that the thoughts, facts, and words are your own unless you say otherwise. Another good reason to cite is that literature citations make your report look more scholarly.

Whenever you take information from another source you must give credit. If you have several ideas together from one source, you need cite the source only once. However, some ideas are so well known that no citation is necessary (e.g., 'Plants are photosynthetic.'); your opening sentence might be sufficiently general (see Appendix). If in doubt, cite a source.

You should paraphrase the original – rewrite it in your own words. There is really no reason to quote from scientific writing, because the writing is secondary to the information and not literary.

There are several methods for citing literature, and you must follow one method consistently. The commonest and easiest is the author-date system: Brown (1990). Some journals save space by using the number method, numbering references either consecutively through the text or according to the alphabetical listing in the References. (See examples below.)

Whatever method you use (your professor may have specified; if not we suggest you use author-date), you must keep the flow of the text smooth. Citations, especially several author-dates in a row, tend to break up the text. Put them at the ends of sentences, or at least at the ends of clauses. Don't break up clauses unless you absolutely must.

You can either use the author's name as part of the sentence – 'Brown (1987) showed that . . . ' – or, usually more concisely, put both name and date in parentheses – 'Cold dogs have been known to eat hotdogs (Brown & White 1988).' If you have more than one paper by the same author(s) in one year, distinguish them by 'a' and 'b' (Jones 1979a). Don't include the author's initials in the citation. (The exception is when you need to distinguish two authors with the same name and publication date, e.g., M. Smith 1990, K. Smith 1990.)

For two authors, you can use 'and' or '&' between them. For a lab report the choice is trivial; just be consistent. Also be consistent about whether or not you put a comma before the date: (Smith, 1976) or (Smith 1976). If a paper has three or more authors, abbreviate using et al. (short for *et alii*, which means 'and others'): Jones et al. (1989). In current style, et al. is not underlined (nor printed in italics).

You should not cite the page number (e.g., Jimenez 1985: 255), in contrast to practice in other disciplines. (The one exception is when you use a direct quotation.)

What do you do about "second-hand" references? When you read textbooks, you will find that most of the facts and ideas there are credited to someone else. Whom do you cite? Cite only the books or papers that you see. You can assume that these published works are authoritative and you can cite them as your sources. Do not include in your reference list works you have not actually consulted.

Here are some examples of good citation style:

'Respiration rates and glycolytic enzyme activites are affected by temperature (Jones 1980, Brown 1986).'
(Two or more references are given in chronological order.)

'Whereas Huff (1960) reported respiratory rates of less than x, more recent studies have consistently found much higher rates (Puff 1987, Wheeze & Gasp 1988).'

If you have to use the number system with alphabetical references, draft your paper using author-date, complete and number your reference list, then substitute numbers for the author-date citations in the text of the *final* draft:

'Whereas Jones (6) reported . . . recent studies have consistently found much higher rates (10, 12).'

'Writer's block in student authors is prevented by Lobban & Schefter's book (5).'

Do not put references in footnotes; collect them in the References section. Whereas footnotes are usual in the humanities, they are unacceptable in science reports. (Footnotes are not used for parenthetic comments, either. Such asides are simply written in parentheses, like this.) In published papers there may be a cluster of footnotes on the first page, giving authors' new addresses, or a list of abbreviations (see Appendix). Some chemistry and biochemistry journals put the references in footnotes, but this is particular journal style. You won't need any footnotes.

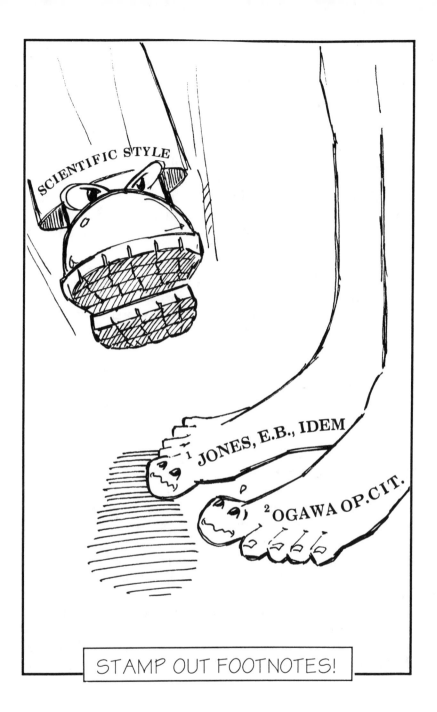

Finding the information for the reference list

Some minor problems may await you as you create your References section. The first is how to write out the title of an article. Capitalize only the first word of the title and any proper names (including genus names). Unfortunately, many journals capitalize every major word, or use block capitals, and may not use italics for Latin names. For instance, here's how the title and authors' names of the paper given in the example on page 64 appeared in the journal:

MUSCLE AND LIVER INTRACELLULAR ACID-BASE AND METABOLITE
STATUS AFTER STRENUOUS ACTIVITY IN THE INACTIVE,
BENTHIC STARRY FLOUNDER PLATICHTHYS STELLATUS[1]

C. LOUISE MILLIGAN AND CHRIS M. WOOD

In this case, you need to notice the Latin name, and you must abbreviate the authors' given names. The superscript ([1]) is not part of the title. The date and citation for this paper appeared at the bottom of the page:

Physiol. Zool. 60(1):54–68. 1987.

Other journals put them at the top of the page or below the abstract. Many journals, such as *Physiological Zoology*, give you their proper abbreviation. The issue number can be dropped. Some journals do not print the page range on the first page (and sometimes there may not even be a number on that page), so you have to flip the pages to find them. If you have only a photocopy of such a first page, you're in for another trip to the library!

You should use the words and punctuation *exactly* as they are in the original title, including any mistakes, except that you may have to insert a colon between a main title and subtitle. If you spot a mistake in the title and want to show that the error is not yours in recopying, you can write [*sic*] after the mistake. (*Sic* is Latin for 'thus'.) For instance:

Sparrow, G. B. & A. Swanson. 1942. Observations on birds: fathers [*sic*] are essential to flight. The Pterodactyl 3: 6–8.

Dates of books are to be found on the back of the title page. Use the date of first publication of that edition, not the date of the most recent reprinting. The copyright date may help, but copyright may be renewed for the reprint. For example, *Scientific Writing for Graduate Students* (Woodford 1968) is "Copyright 1968, 1976, 1981, 1983, and 1986." Yet, it is still the first edition. Its printing history is given as:

First edition 1968
First reprinting 1976
Second reprinting 1981
Third reprinting 1983
Fourth reprinting, with references updated by F. Peter Woodford, 1986.

Large publishing houses have offices worldwide and list them all on the title page (see ours). Pick the first name in the list (probably the head office) or the one in your country.

Some series, although edited books, are sufficiently well known or established that you can cite them as if they were journals. (Of course, journals have editors too, but they never appear in the references.) Examples of such series are titles beginning *Annual Review of* and *Advances in*. If there is a volume number (and the volume is not simply a part work), you can treat it as a journal.

Format for references

For your lab report it may be enough if you have all the necessary information in a suitable and consistent style. Journals, on the other hand, are very particular about reference style, down to the last comma and period. If your instructor wishes you to follow a certain journal style, she or he will tell you. Follow that style exactly.

In addition to alphabetizing by first author, use the following rules for arranging the references:

More than one paper by same author:
    put in chronological order
More than one paper in same year by same author:
    number a, b (the order doesn't matter, just be sure the letters correspond to the citations in the text)
Dual author names with same first author:
    alphabetize according to second author
More than one paper by author et al., even if the coauthors are different:
    put in chronological order regardless of et al.'s names
More than one author with the same last name:
    alphabetize according to initials

The following sample list is correctly arranged according to these rules:

Adams, K. 1978.
Adams, K. 1980a.
Adams, K. 1980b.
Adams, N. 1940.
Adams, N. & B. Smyth. 1989.
Adams, N. & K. Wong. 1986.
Adams, N., L. Brown & B. Smyth. 1984.
    [= Adams et al. 1984]
Adams, N., K. Adams & Z. Adams. 1986.
    [= Adams et al. 1986]

The information needed for complete references depends on whether the source is a journal article, a chapter in a book, or a whole book. See examples at the beginning of this chapter, in the Appendix, and in the list of Further Reading.

For a journal article you need (in this order):

Names of all authors (family names and initials; don't write out given names)
Date published
Title of paper (do not put in quotation marks)
Journal name (usually abbreviated and sometimes underlined [Ask!])
Volume number
[Issue number only if issues have separate pagination (as in *Scientific American*)]
First and last page numbers
[Very few biological journals put the date at the end of the reference, whereas many humanities journals do.]

For a book you need:

Names of all authors or editors
Date this edition originally published (don't count reprintings)
Title of book (may be underlined)
[Volume number if the work is in parts]
Edition number (if not the first)
Name and city of publisher (if city is not well known also give state or country)
Total number of pages in book (not essential)

For a chapter in a book or a paper in conference proceedings, you need:

Names of all authors
Date
Title of article/chapter

Part II: Style

The word, In
Names of all editors
Title of book
First and last pages of article
Name and city of publisher
[The last five elements are often arranged in some other order.]

The order of names and initials varies from journal to journal. The first author's initials are always placed after the name so that alphabetization is easy. Some journals have all names this way, others prefer to have remaining names and initials in the normal order, initials first:

Jones, K. & Brown, J. 1980.
or
Jones, K. & J. Brown. 1980.

The punctuation and connectors between names also vary. Some journals use 'and', some use '&', and some just use a comma. You should simply be consistent. Similarly, the punctuation around the date varies.

Chapter 11   Sample Lab Report and Revision

Style – in scientific writing or any other – is not so much a matter of following a set of rules, as of writing, being corrected, looking at the suggestions for improvement, and writing again until you can become your own editor. It is a gradual process of improvement. The best way we can illustrate style is to present a sample 'student' lab report, thoroughly annotated by 'the professor', and a revised report. If the first version seems worse than you could write for your first report, that's good! We wanted to illustrate plenty of common mistakes. The revised report is much better but could still be improved; it is not the best that A. Student could write. We suggest you criticize the revised version in the way that we edited the draft, as an exercise to help you revise your own reports. (Criticizing someone else's work is much easier than taking a cold, hard look at your own writing!)

The following standard proofreaders' marks were used in editing the sample report:

⧣	insert space	⊆	use capital letter
ℓ	delete	ꟷꟷ	indent
ℓ̸	delete and close up	⌐	transpose
⊙	period	⁋	new paragraph

Part II: Style

Antibiotic Sensitivity Testing

by A. Student

Introduction

When a physician prescribes treatment for an
infectious disease, (s)he has to know which antibiotic   ①
will be most effective against the causative pathogen.
Modern medicine <u>made great strides</u> after Alexander   ②
Fleming accidentally discovered penicillin in 1928.
Synthetic penicillin was developed during World War II
and soon other antibiotics were discovered (Encycl.   ③
Brit. 10:990). These compounds ~~work by~~ killing or
inhibiting the growth of disease-causing microorganisms.

① Structure: put this sentence in when you
    introduce the problems. Better to start
    with Fleming's discovery.

② Cliché; informal.

③ What is this reference?! You have to find
    author and date.

74

Sample lab report and revision

A. Student: Antibiotic Testing, page 2

Chemotherapeutic agents, like "magic bullets," *(4)*
selectively attack the unwanted microorganisms without *(5)*
harming the human cells *(6)* (Brock et al., 1984).

Unfortunately problems arose with these wonder
cures (Benson, 1985). Some germs developed resistance. *(7)*
Scientists are now searching for new antibiotics
against malaria and denge fever, for instance, because
antibiotic-resistant mutants have appeared.

no ¶ Another problem is that some supposed cures did
not work. Some antibiotical agents affect only some *(8)*
organisms. This is called selective toxicity.

Lab technicians can easily test the effectiveness
of an antibiotic using simple media and prepared
antibiotic disks *(6)* (Brock et al. 1984). An antibiotic *(6)*
disk is put on a lawn of a pure bacteria culture in a
petri dish. The area around the disk where the bacteria

(4) Misleading; delete.
(5) Single-sentence paragraph; merge with preceding.
(6) Citation goes at end of sentence, before period.
(7) Colloquial.
(8) Wordy!

Part II: Style

do not grow is termed the zone of inhibition, if there is any. However interpreting the results was a problem. The zone of inhibition varies with the diffusibility of the agent, the size of the inoculum, the type of medium, and many other factors. The Kirby-Bauer team solved the problem of what the results meant and their method is widely approved today.

To demonstrate that some antibiotics are more effective than others against different organisms, we tested eight commonly used antibiotics, including penicillin-G, against three laboratory cultures of common bacteria. The bacteria were chosen to represent three general types of cells: a Gram-positive coccus (Micrococcus), a Gram-positive rod (Bacillus), and a Gram-negative rod (Escherichia coli).

⑨ Smoother and more concise put together:
 'Interpretation was a problem until Kirby &
 Bauer suggested a standardized technique.'

⑩ Dangling infinitive.

⑪ Objectives are not defined clearly enough.

⑫ Introduce the antibiotics more; perhaps
 also the bacteria (medically important?).

A. Student: Antibiotic Testing, page **4**

Methods

The procedure used was similar to the Kirby-Bauer method (Benson, 1985). Variations are mentioned here. (13)
Petri dishes were prepared with (14) (TSA) agar having a pH of (15) 7.2-7.4 and a uniform thickness of 4 mm.   Then the dishes were labeled with the first letter of the genus of the organism to be tested. Each petri dish was (16) swabbed with a broth of <u>Bacillus</u> sp., <u>Escherichia coli</u>, or <u>Micrococcus</u> sp.   They were all left to dry for 3-5 minutes. Antibiotics tested were (AM-10) Ampicillin, (C-30) (17) Chloramphenicol, (E-15) Erythromycin, (GM-10) Gentamicin, (NA-30) Nalidixic Acid, (P-10) Penicillin G, (PB-300) Polymyxin B, and (S-10) Streptomycin. Then a set of the eight antimicrobial disks were randomly applied with

(13) Delete: you have written your procedures below, rather than simply noting variations from the standard K-B.

(14) Explain abbreviation.

(15) Too much detail.

(16) 'Agar surfaces were...'

(17) Leave out manufacturer's code numbers.

Part II: Style

A. Student: Antibiotic Testing, page 5

the BBL dispenser to each dish. Glass covers were ⑮
replaced, the dishes inverted, and incubated for 48
hours at 37° C.　└ Plates were

After incubation, ~~keeping the dishes covered~~ the
diameter of the zone of inhibition around each coded ⑱
disk was measured (without magnification) by pressing
the metric ruler to the outside glass and reading to
the nearest millimeter. Faint growth and tiny colonies
that could only be detected by close scrutiny were
ignored.

To determine the degree of relative sensitivity
(resistance or sensitivity) of an organism to each ⑲
antibiotic, the Kirby-Bauer chart Appendix A was
referenced (Benson, 1985).

⑱ Wow! Dangling and otherwise awkward
    phrasing; long & convoluted sentence.
    Best way to fix is to delete unnecessary
    details and use active voice.
⑲ More dangling & awkward phrases.

78

A. Student: Antibiotic Testing, page 6

Results

Six of the eight antibiotics clearly inhibited at least one of the bacteria (Table 1). Of these, only three inhibited both E. coli and Bacillus; these were chloramphenicol, gentamicin and streptomycin. Penicillin G did not kill either of these species, and the effects of nalidixic acid were intermediate.

All the bacteria were sensitive to at least one antibiotic. E. coli and Bacillus were each sensitive to three compounds. Around Micrococcus the zones of inhibition completely overlapped so that effects of individual compounds could not be distinguished.

(20) Results are presented in two different ways; this is confusing. Clarify your objectives (11) and then you should be able to clarify Results presentation.

(21) 'Inhibit' — penicillin does not kill the bacteria but prevents growth. See note (12).

(22) So, obviously it was very sensitive!

Part II: Style

A. Student: Antibiotic Testing, page 7

Discussion

The results show that antibiotics are effective against some species and not others. Although results were as expected, there were some sources of error. ㉓

The <u>Micrococcus</u> data could not be used because the zones merged. Obviously, <u>this bacterium is very</u> ㉔ sensitive but I could not compare data with the Kirby-Bauer table. This problem could be remedied by having more space between disks.

Other measurement errors arose when the zones of inhibition were not so neatly circular. This happened for example when zones were irregularly shaped or when two or more zones merged. When the antibiotic disk was placed too near the edge of the petri plate the zone of inhibition was semicircular. Due to the small sample

㉓ a. You have hardly discussed the results before launching into the errors.

b. First sentence is an attempt to start with a positive statement about the results, but in fact is so general that it merely repeats the Introduction. Be specific.

㉔ See note ㉒. You are discussing a result which was not given in Results.

80

Sample lab report and revision

A. Student: Antibiotic Testing, page 8

size, we, interpolated the other half of the diameter so we could have some data.

Another possible source of error was introduced when a similar-looking organism grew within the zone of inhibition. Sometimes these extraneous colonies were of a different shape and color and size from the original organism tested so were apparently resistant mutants or contaminants. For our measurement, they were ignored.

Certain antibiotics are apparently effective against more disease-causing organisms. These antibiotics are termed broad-spectrum antibiotics. Others are effective against only certain bacteria. These are termed narrow-spectrum. Some disease-causing organisms are apparently resistant to many antibiotics. On the basis of this study, C-30 and GM-10 could be called broad spectrum antibiotics.

㉕ So these are _not_ sources of error!

㉖ These conclusions are too broad, based on so few tests. Review objectives.

㉗ Don't use the royal 'we' when you mean 'I.'

㉘ Give names of compounds.

Part II: Style

A. Student: Antibiotic Testing, page 9

References

Benson, Harold J. 1985. Microbiological Applications: a Laboratory Manual in General Microbiology - 4th ed. ~~Complete version.~~ *Publisher, place.*

Brock, ~~Thomas~~ D., ~~David~~ W. Smith, and ~~Michael~~ T. Madigan. 1984. Biology of Microorganisms, 4th ed. Prentice-Hall, *Englewood Cliffs,* N.J.

Author(s), date
Encyclopaedia Britannica. Antibiotic. *which edition?* vol. 1, pp. 986-990. *Publisher*

29 Give initials only
30 Indent second and subsequent lines

82

A. Student: Antibiotic Testing, page 10

Table 1. Results of tests. Zone widths *[handwritten: Inhibition diameters]* /in mm and resistance (R) or sensitivity (S).
[handwritten: say what]

BACTERIA	ANTIBIOTICS							
	AM-10	C-30	E-15	GM-10	NA-30	P-10	PB-300	S-10
Bacillus sp.	11 (R)	24½ (S)	25 (S)	18 (S)	18 (I)	8 (R)	18 (S)	14 (I)
	10 (R)	23 (S)	30 (S)	22 (S)	15 (I)	8 (R)	8 (R)	20 (S)
	11 (R)	24 (S)	30 (S)	21 (S)	--	8 (R)	8 (R)	22 (S)
E. coli	19 (S)	23 (S)	18 (S)	20 (S)	8 (R)	11 (R)	15 (S)	16 (S)
	18 (S)	30 (S)	11 (I)	18 (S)	20 (S)	20 (R)	12 (S)	17 (S)
	20 (S)	30* (S)	15 (I)	10 (S)	14* (I)	7 (R)	11* (I)	8* (R)

*irregular shape *[handwritten: of ∧?]*

[handwritten annotations:]

③ Use antibiotic names

㉜ Raw data! The chart in the labs manual is a good way to record your results but not to present them in report. Give mean values.

㉝ Try arranging table with species across the top.

83

Part II: Style

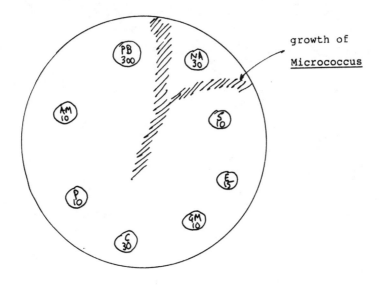

growth of
Micrococcus

Figure 1. Drawing of agar plate showing overlapping zones of inhibition of **Micrococcus** from eight antibiotics.

Is this drawing really useful? — The statement in the text says it all.

Sample lab report and revision

Antibiotic Sensitivity Testing

(Revision)

by A. Student

Introduction

Alexander Fleming's accidental discovery of penicillin in 1928 led to major advances in treatment of diseases. Synthetic penicillin was developed during World War II and soon other antibiotics were discovered (Weinstein, 1980). These compounds kill or inhibit the growth of disease-causing microorganisms without harming human cells (Brock et al., 1984). Unfortunately, problems arose with these wonder cures (Benson, 1985). Some germs developed resistance and other organisms were not affected at all by them. However, physicians must know which antibiotic will be most effective against a pathogen.

Lab technicians can easily test the effectiveness of an antibiotic using simple media and prepared antibiotic disks (Brock et al., 1984). An antibiotic disk is put on a lawn of pure bacterial culture in a petri dish. The area around the disk where the bacteria do not grow is termed the zone of inhibition, if there is any. However, the zone of inhibition varies with the diffusibility of the agent, the size of the inoculum,

85

Part II: Style

A. Student
Antibiotic Testing (Revised), page 2

the type of medium, and many other factors. Interpretation was a problem until Kirby & Bauer suggested a standardized technique (Benson, 1985).

We tested eight common antibiotics, including penicillin, against three different types of bacteria. The organisms included a Gram-negative rod (<u>Escherichia coli</u>), a Gram-positive rod (<u>Bacillus</u> sp.), and a Gram-positive coccus (<u>Micrococcus</u>). E. coli is the coliform bacterium that is counted in public waters as a measure of fecal contamination, and the other two species live on skin. <u>Micrococcus</u> is similar to <u>Staphylococcus</u> that is a major source of infections in hospitals (Brock et al., 1984).

The antibiotics included some that act on certain bacterial walls (penicillin, ampicillin), some that interfere with protein synthesis (chloramphenicol, streptomycin, erythromycin, gentamicin), and one that acts on cell membranes (polymixin). Some are broad-spectrum antibiotics, toxic to most species (e.g., compounds that affect protein synthesis), others are narrow-spectrum (e.g., penicillin and ampicillin, that affect mainly Gram-positive bacteria) (Weinstein, 1980).

86

Sample lab report and revision

A. Student
Antibiotic Testing (Revised), page 3

The objective was to show that a given strain of bacterium is sensitive to some antibiotics but resistant to others.

Methods

The procedure used was similar to the Kirby-Bauer method (Benson, 1985). Petri dishes had been prepared with Tryptic Soy Agar (TSA, Bacto-Difco). I swabbed each agar plate with a broth of one of the three bacteria. When the surface was dry I applied a set of eight antibiotic disks randomly on each plate, using a disk dispenser. The antibiotics are listed in Table 1. Plates were incubated for 48 hours at 37° C. I measured the diameter of the zone of inhibition around each disk to the nearest millimeter, ignoring faint growth and tiny colonies that could be detected only by close scrutiny. Resistance or sensitivity was determined by comparing the zone diameters with those in the Kirby-Bauer table (Benson, 1985).

Part II: Style

A. Student
Antibiotic Testing (Revised), page 4

Results

Escherichia coli was resistant only to penicillin
G, although results for erythromycin and nalidixic acid
were intermediate (Table 1). Bacillus sp. was resistant
to four of the eight antibiotics and resistant to the
others. Zones of inhibition around disks on Micrococcus
merged and could not be measured, but this bacterium
was evidently sensitive to most of the antibiotics.

Discussion

Penicillin G inhibits wall synthesis in Gram-
positive bacteria and was not expected to affect the
Gram-negative rod, E. coli. The resistance of the
Bacillus to this antibiotic is surprising. This
bacterium was freshly isolated and may have plasmids
with genes for penicillin resistance (Prescott et al.,
1990). This illustrates the difficulty in prescribing
an antibiotic for a given infection.

Both bacteria were sensitive to the four antibiotics that inhibit protein synthesis (chloramphenicol, erythromycin, gentamicin, and streptomycin);
these are broad-spectrum antibiotics (Weinstein, 1980).

Sample lab report and revision

Table 1. Effects of antimicrobial compounds against two
bacteria in the Kirby-Bauer test.

Antibiotic	Bacillus sp.		E. coli	
	zone [a]	sensit. [b]	zone	sensit.
Ampicillin	11±1	R	19±1	S
Chloramphenicol	24±1	S	28±4	S
Erythromycin	28±3	S	15±4	I
Gentamicin	20±2	S	19±1	S
Nalidixic acid	17±2	R	14±6	I
Penicillin G	8±0	R	13±7	R
Polymixin B	11±6	R	13±2	S
Streptomycin	19±4	S	14±5	S

[a] Mean width of inhibition zones to nearest mm ± SD.
[b] Sensitivity based on comparison with Kirby-Bauer chart
(Benson, 1985): R = resistant, S = sensitive, I = intermediate.

Part II: Style

Some combinations gave inconsistent results. For instance, <u>Bacillus</u> showed resistance to polymixin in two dishes but sensitivity in the third. The sensitive one may be a revertant mutant (Prescott et al., 1990).

The Gram-positive <u>Micrococcus</u> appeared to be sensitive to all the antibiotics, but I cannot draw any conclusion because the widths of zones of inhibition must be compared with the numbers in the Kirby-Bauer table. Overlap of zones could be prevented with more space between disks, or by putting each disk on a separate plate of the disease organism.

Bacteria respond differently to different antibiotics and may lose or acquire resistance via plasmids and mutation. The Kirby-Bauer method provides a rapid test for the sensitivity of a particular strain of pathogen to different antibiotics. It gives important information to the physician who must prescribe an antibiotic for a patient.

Sample lab report and revision

A. Student
Antibiotic Testing (Revised), page 7

References

Benson, H.J. 1985. Microbiological Applications: A Laboratory Manual in General Microbiology, 4th ed. Wm. C. Brown, Dubuque, Iowa.

Brock, T.D., D.W. Smith & M.T. Madigan. 1984. Biology of Microorganisms, 4th ed. Prentice-Hall, Englewood Cliffs, N.J.

Prescott, L.M., J.P. Harley & D.A. Klein. 1990. Microbiology. Wm. C. Brown, Dubuque, IA.

Weinstein, L. 1980. Antibiotic. Encyclopaedia Britannica, 15th ed., Macropaedia, vol. 1, pp. 986-990. University of Chicago Press, Chicago.

Appendix The Anatomy of a Scientific Paper

The following annotated excerpts from a published scientific paper are intended primarily to show you the *format* of a report. As you develop your science writing skills, you can also study the writing style.

MUSCLE AND LIVER INTRACELLULAR ACID-BASE AND METABOLITE STATUS AFTER STRENUOUS ACTIVITY IN THE INACTIVE, BENTHIC STARRY FLOUNDER PLATICHTHYS STELLATUS[1]

C. LOUISE MILLIGAN AND CHRIS M. WOOD

Friday Harbor Laboratories, University of Washington, Friday Harbor, Washington 98250; and
[2]Department of Biology, McMaster University, Hamilton, Ontario, Canada L8S 4K1

(Accepted 6/18/86)

In addition to an extracellular acidosis in which blood metabolic acid load greatly exceeded lactate load, exhaustive activity in starry flounder resulted in an intracellular acidosis of largely metabolic origin in the white muscle, with intracellular pH dropping from 7.56 to 7.27, as measured by DMO distribution. An accumulation of lactate and depletion of glycogen in addition to a shift of fluid from the extracellular to intracellular space were associated with the postexercise acidosis. Pyruvate levels increased in blood and later in muscle; the relative rise in pyruvate was greater than that in lactate so the lactate:pyruvate ratio declined. The muscle intracellular acidosis was corrected sooner than the extracellular acidosis (4–8 h vs. 8–12 h). The restoration of muscle pHi was associated with an increase in pyruvate, a restoration of glycogen stores, and clearance of the lactate load. It is suggested that both lactate and acidic equivalents (H$^+$) were cleared from the muscle via in situ oxidation and/or glyconeogenesis and that the rapid correction of the intracellular acidosis through efflux of part of the H$^+$ load facilitated metabolic recovery. The liver showed a progressive alkalinization after exercise. This alkalinization was of metabolic origin and not associated with lactate accumulation. Except for a short-lived depression 0.5 h after exercise, red cell intracellular pH remained virtually constant.

INTRODUCTION

Intense activity in vertebrates results in the accumulation of lactate and acidic equivalents (H$^+$) in the working m̄ ̄ ̄ Both end pr̄ ̄ ̄
s̄p̄ ̄

then enters the blood to be taken up by the muscle and utilized to replenish ·'·· ·ˑ· stores (Newsholme ȃr̄ ̄ ̄ ̄

ABSTRACT
(not usually headed)
Notice the specific facts. There are no references to literature or authors' graphs/tables. This abstract focuses exclusively on results.

[1] We thank the director, Dr. A. O. D. Willows, and staff of Friday Harbor Laboratories, University of Washington, for their assistance and hospitality. This work was funded by an NSERC operating grant to C.M.W. C.L.M. was supported by an Ontario Graduate Scholarship.
[2] Permanent address and address for correspondence and reprints.

Footnotes are used for current addresses and, in this journal, for acknowledgments.

Physiol. Zool. 60(1):54–68. 1987.

Excerpts reprinted with permission of the publisher and authors.

INTRODUCTION

A general statement starts the article; no reference is needed.

Notice citation style

This is the hypothesis (which they will disprove) – see DISCUSSION.

Sets up the study question

This paragraph gives the objectives.

Significance of the study

Intense activity in vertebrates results in the accumulation of lactate and acidic equivalents (H^+) in the working muscle. Both end products appear in the blood space, though, in flounder, lactate does not accumulate to any great extent (1–2 mmol/liter), with blood levels only one-quarter those of H^+ (Wood, McMahon, and McDonald 1977; Milligan and Wood 1987). The classical picture is the Cori cycle: lactate and H^+ leave the muscle and are transported via the blood, to the liver, where they are converted to glucose. The glucose then enters the blood to be taken up by the muscle and utilized to replenish glycogen stores (Newsholme and Leech 1983). However, the fates of lactate and H^+ after exercise in relatively inactive, benthic fish such as starry flounder are unclear. The data of Wardle (1978) and Batty and Wardle (1979) suggest that lactate, and presumably H^+, are not transported out of the muscle but rather are utilized as substrates for in situ glyconeogenesis.

The present study investigates the possible fates of H^+ and lactate after exercise in starry flounder. Using the DMO method for measuring intracellular pH (pHi; Waddell and Butler 1959), we examined the intracellular acid-base and metabolite changes associated with exercise in the white-muscle mass, liver, and blood. In addition, we investigated the effect of the exercise-induced extracellular acidosis on red blood cell (RBC) pHi. This study complements our work on the active, pelagic, rainbow trout (Milligan and Wood 1986b) and thus provides insight into the physiological reason(s) for the species-dependent pattern of lactate and H^+ accumulation in the blood.

MATERIALS AND METHODS

Notice how reference to published procedures is handled.

The section can be divided into titled subsections, if need be.

RESULTS

A statement of the result is made with reference to a table.

Results and Discussion can also be subdivided in long papers.

MATERIAL AND METHODS

EXPERIMENTAL ANIMALS

Starry flounder were captured, held at 11 ± 1 C, and catheterized in the caudal artery as described in the companion paper (Milligan and Wood 1987).

EXPERIMENTAL PROTOCOL

In the present experiments, fish were sampled only once (terminally) rather than sequentially for analysis of blood, muscle, and liver acid-base and metabolite status. Approximately 12 h prior to sampling were infused with
ml ^{14}C

...central ... from the respective postexercise mean concentration at each sample time.

STATISTICAL ANALYSIS

Means ± 1 SEM (n) are reported throughout, unless stated otherwise. Differences between groups were tested for significance ($P < .05$) with Student's two-tailed t-test, unpaired design.

RESULTS

TISSUE BUFFER CAPACITIES

The buffer capacity of muscle was ~ 1.6 times that of liver (table 1). These values represent total physicochemical buffer capacity (i.e., nonbicarbonate + bicarbonate). However, since PCO_2 was kept low during titration, intracellular [HCO_3^-] would be very low (<1 mmol/liter) and would not contribute significantly to the measured β value.

EXTRACELLULAR ACID-BASE, METABOLITE, AND ELECTROLYTE STATUS

Changes in hematology and extracellular (i.e., plasma) arterial pH (pHa), $PaCO_2$, and

An important part of this paper was the appropriateness of the techniques, so the discussion clears these up before going into the results per se.

Discussion of sources of error

Sometimes techniques don't work.

DISCUSSION

METHODOLOGY

The potential problem of DMO disequilibrium between ICFV and ECFV in the dynamic postexercise situation has been considered and discounted in the companion paper (Milligan and Wood 1987). A second potential source of error in the pHi estimate is the assumption that plasma pHa is representative of the ECF. The white-muscle mass is perfused with both arterial and venous blood, with the true interstitial pH lying between pHa and venous pH (pHv). The teleost liver, however, is perfused mainly by venous blood (Smith and Bell 1976). However, at least in rainbow trout, we have shown that arterial versus venous pH, [DMO], and [mannitol] differences were insignificant except immediately after exercise and that even then they tended to self-compensate, thus having negligible effect on calculated whole-body and tissue pHi (Milligan and Wood 1986a, 1986b).

Use of mannitol for an ECFV marker in flounder liver proved unsuccessful. A similar difficulty was encountered in estimating liver and heart ECFV with mannitol in rainbow trout (Milligan and Wood 1986b). inulin-derived estimates for liver ECFV flounder were similar to those 'tribution (C. M. d data)

The authors' results are
compared to published
values on similar species (other
fish) . . .

RBCs had the lowest rest pHi of the tissues examined, similar to pHi values reported for other fish, such as the rainbow trout (Milligan and Wood 1986a, 1986b; Primmet et al. 1986).

Flounder white-muscle pHi (fig. 2) was 0.2–0.3 units greater than values reported for white muscle from more active pelagic species, such as rainbow trout (Hōbe, Wood, and Wheatly 1984; Milligan and Wood 1986b), dogfish (Heisler, Weitz, and Weitz 1976), and eel (Walsh and Moon 1982). However, it was not dissimilar to values reported for the relatively inactive, demersal catfish (Cameron and Kormanik 1982) and sea raven (Milligan and Farrell 1986).

. . . and the differences are
discussed.

These interspecies differences may be related to muscle lactate levels, for the active species with the lower pHi values generally tended to have higher muscle lactate levels (8–10 mmol/kg vs. 1–3 mmol/kg).

. . . . .ued lac-
. . . . lactate was retained, as suggested by Wardle (1978) and Batty and Wardle (1979).

The results are related back to
the hypothesis; data supported
the *null* hypothesis.

We suggest that most of the lactate and the greater proportion of the H$^+$ remained in the white-muscle mass and were metabolized in situ, i.e., that the Cori cycle was of little importance. The partial efflux of H$^+$ in combination with this metabolism would explain the rapid correction of muscle pHi and subsequent intracellular alkalosis (Milligan and Wood 1987). Fish white muscle is capable of lactate oxidation, albeit at lower rates than aerobic tissues (e.g., heart and red muscle; Bilinski and Jonas 1972). For flounder muscle to clear ΔLa$^-$ via in situ metabolism would require average O$_2$ cons⸱⸱⸱⸱⸱⸱⸱⸱⸱ ⸱ ⸱
⸱⸱ /¹

pattern

acetate clearance was ob-
after exercise in the sluggish sea
raven (Milligan and Farrell 1986). We suggest that restoration of both glycogen stores
and the potential for further metabolism in
white muscle is of prime importance in the
relatively inactive species and that rapid
correction of the intracellular acidosis in
preference to the extracellular acidosis fa-
cilitates this process. The fish's survival depends more on the ability to use the muscle
again glycolytically as soon as possible
rather than on the capacity for aerobic respiration and blood O_2 transport.

Finally, the results are related
back to the wider realm of fish
biology.

= REFERENCES

LITERATURE CITED

ABE, H., G. P. DOBSON, U. HOEGER, and W. S. PARK-
HOUSE. 1985. Role of histidine-related compounds
in intracellular buffering in fish skeletal muscle. Am.
J. Physiol. 249:R449–R454.

ADCOCK, P. Y., and P. R. DANDO. 1983. White muscle
lactate and pyruvate concentrations in rested
flounder, Platichthys flesus, and plaice, Pleuronectes
platessa: a re-evaluation of handling and sampling
techniques. J. Mar. Biol. Assoc. UK 63:897–903.

BATTY, P. S., and C. S. WARDLE. 1979. Restoration
of glycogen from lactic acid in th

lactate in frog and rabbit muscle. Biochem. J. 118:
887–893.

BILINSKI, E., and R. E. E. JONAS. 1972. Oxidation of
lactate to carbon dioxide by rainbow trout (Salmo
gairdneri) tissues. J. Fisheries Res. Board Can. 29:
1467–1471.

CAMERON, J. N., and G. A. KORMANIK. 1982. Intra-
and extracellular acid-base status as a function of
temperature in the freshwater cha

Second and subsequent lines
indented

Chapter in a book

1972. Comparative studies on the metabolism of shallow-water and deep-sea marine
fishes. I. White-muscle metabolism in shallow-water
fishes. Mar. Biol. 13:222–237.

HASSID, W. Z., and S. ABRAHAM. 1957. Chemical
procedures for analysis of polysaccharaides. Pages
34–37 in S. P. COLOWICK and N. O. KAPLAN, eds.
Methods in enzymology. Vol. 3. Academic Press,
New York.

HEISLER, N., and P. NEUMANN. 1980. The role of
physico-chemical buffering and of bicarbonate
transfer processes in intracellular pH regulation in
response to changes in temperature in the larger-
spotted dogfish (Scyliorhinus stellaris). J. Exp. Biol.
85:99–110.

HEISLER, N., H. WEITZ, and A. M. WEITZ 1976. I
tracellular and

cellular and extracellular
acid-base status and H^+ exchange with the environment in the inactive, benthic starry flounder
Platichthys stellatus. Physiol. Zool. 60:37–53.

———. 1986a. Intracellular and extracellular acid-base
status and H^+ exchange with the environment after
exhaustive exercise in the rainbow trout. J. Exp.
Biol. 123:93–121.

———. 1986b. Tissue intracellular acid-base status and
the fate of lactate after exhaustive exercise in the
rainbow trout. J. Exp. Biol. 123:123–144.

NEWSHOLME, E. A., and A. R. LEECH. 1983. Biochemistry for the medical sciences. Wiley, Chi-
chester. 952 pp.

NIKINMAA, M. 1982. Effects of adrenaline on red cell
volume and concentration gradient of meta-
across the red cell memb

A book

Further Reading

Writing in general [only a sample]

Baker, S. 1989. The Practical Stylist, 7th ed. Harper & Row, New York.

Gowers, E. 1990. The Complete Plain Words, revised by S. Greenbaum & J. Whitcut. D.R. Godine Publ., Boston.

Leggett, G.H., C.D. Mead & W. Charvat. 1988. Prentice-Hall Handbook for Writers. Prentice-Hall, Englewood Cliffs, N.J.

Strunk, W., Jr. & E.B. White. 1979. The Elements of Style, 3rd. ed. Macmillan, New York.

Turabian, K.L. 1973. A Manual for Writers of Term Papers, Theses, and Dissertations, 4th ed. U. Chicago Press, Chicago.

Turabian, K.L. 1977. Student's Guide for Writing College Papers, 3rd ed. U. Chicago Press, Chicago.

University of Chicago Press. 1982. The Chicago Manual of Style, 13th ed. U. Chicago Press, Chicago.

Scientific writing

Council of Biology Editors. 1983. CBE Style Manual, 5th ed. CBE, Bethesda, Md.

Day, R.A. 1988. How to Write and Publish a Scientific Paper, 3rd ed. Oryx Press, Phoenix.

Farr, A.D. 1985. Science Writing for Beginners. Blackwell Scientific, Oxford.

Jones, W.P. & M.L. Keane. 1981. Writing Scientific Papers and Reports, 8th ed. Wm. C. Brown, Dubuque, Ia.

101

Further reading

Stapleton, P. (n.d.) Writing Research Papers: An Easy Guide for Non-Native-English Speakers. Australian Centre for International Agricultural Research, Canberra.

Trelease, S.F. 1958. How to Write Scientific and Technical Papers. Williams & Wilkins, Baltimore. (Paperback edition 1969 by M.I.T. Press, Cambridge, Mass.)

Woodford, F.P. (ed.) 1968. Scientific Writing for Graduate Students. CBE, Bethesda, Md.

Acknowledgments

Much of what C.L. learned about teaching science writing came from using Day's book, supplemented with the CBE books and Stapleton's, all of which are designed for writers of scientific papers for publication. In distilling their ideas and our own experiences with (and *as*) frustrated undergraduate students, we have taken away much but we hope we have added also. The manuscript benefitted greatly from a critical review by Dr. Robert Rasmussen (Humboldt State University), as well as input from Drs. Ernest Matson, Lynn Raulerson, and Eugene Bruce (all colleagues at U. Guam), Monica Ada (a bona fide student at U. Guam), and Dr. Kathleen Zylan, our editor at Cambridge University Press. In the end, of course, the book remains our personal view of how to write a lab report. We appreciate the willingness of authors and publishers who allowed us to use excerpts from their works. Finally, thanks to Jeff Harris (Island Type & Art, Guam) for turning our crude sketches into professional cartoons.

Index

Index